Loutfi Benyettou

DSP processors

Loutfi Benyettou

# DSP processors

## Features, Architectures and Software Development Tools

ScienciaScripts

**Imprint**
Any brand names and product names mentioned in this book are subject to trademark, brand or patent protection and are trademarks or registered trademarks of their respective holders. The use of brand names, product names, common names, trade names, product descriptions etc. even without a particular marking in this work is in no way to be construed to mean that such names may be regarded as unrestricted in respect of trademark and brand protection legislation and could thus be used by anyone.

Cover image: www.ingimage.com

This book is a translation from the original published under ISBN 978-613-8-44965-2.

Publisher:
Sciencia Scripts
is a trademark of
Dodo Books Indian Ocean Ltd. and OmniScriptum S.R.L publishing group

120 High Road, East Finchley, London, N2 9ED, United Kingdom
Str. Armeneasca 28/1, office 1, Chisinau MD-2012, Republic of Moldova, Europe
Printed at: see last page
**ISBN: 978-620-6-29650-8**

# TABLE OF CONTENTS:

## INTRODUCTION

DSPs *(Digital Signal **Processors**) are* processors specifically designed for digital signal processing. The diversity of applications has enabled these devices to evolve considerably since they first appeared in the 1980s. The best-known example is the TMS32010 fixed-point DSP from Texas Instruments, the world market leader in this type of technology. In fact, there are two main families of this type of processor: fixed-point DSPs using integer arithmetic, and floating-point DSPs operating in floating-point arithmetic [1,2]. The latter are the fastest and most accurate, but also the most expensive.

In this course, we present the general characteristics found in most DSPs. The so-called 'conventional' architecture found in first-generation DSPs, and in the vast majority of today's processors, is also presented. The general history of the evolution of these different architectures up to the most recent is outlined. The place occupied by DSPs relative to other computing structures and, finally, the DSP software development tool (composer code).

# CHAPTER 1

## I. HISTORY OF DSPs

For several years now, digital signal processing has been the technique that has attracted the most interest from researchers and specialists. The technology of the processors used in this field, commonly referred to by the acronym DSPs, is still undergoing continuous innovation. Historically, DSPs were initially developed for military and encrypted telecommunications applications in the 1970s. In 1978, Texas Instruments introduced a DSP for voice synthesis dedicated to consumer applications. It took another 15 years for DSPs to become essential components in the vast field of electronics [3].

The evolution of these DSPs has led us to classify them into five main generations according to their performance in various signal processing applications (Figure 1) [4].

- **Generation 1**: Introduction of the Harvard architecture, which separates processing units, addresses and data. The data format is generally 16-bit fixed point.

- **Generation 2**: Technological improvements (CMOS, reduced cycle time); reinforcement of the instruction set to better identify the calculation and addressing primitives linked to the properties of signal processing algorithms, and increased storage capacity.

- **Generation 3**: Data processing in 32-bit floating-point format. Improvement and modification of the architecture of the processing unit by the presence of general registers and the multiplication of transfer paths. Increased complexity reinforced by a significant number of utility peripherals: serial port, Timer, and DMA supported by a clear improvement in multiprocessor communication. This last feature is in line with developments in general-purpose microprocessors.

- **Generation 4**: support for multiprocessor parallelism.

- **Generation 5**: Corresponds to the fixed-point DSPs currently on the market. Compared with the second generation, they are faster, more powerful and have more memory.

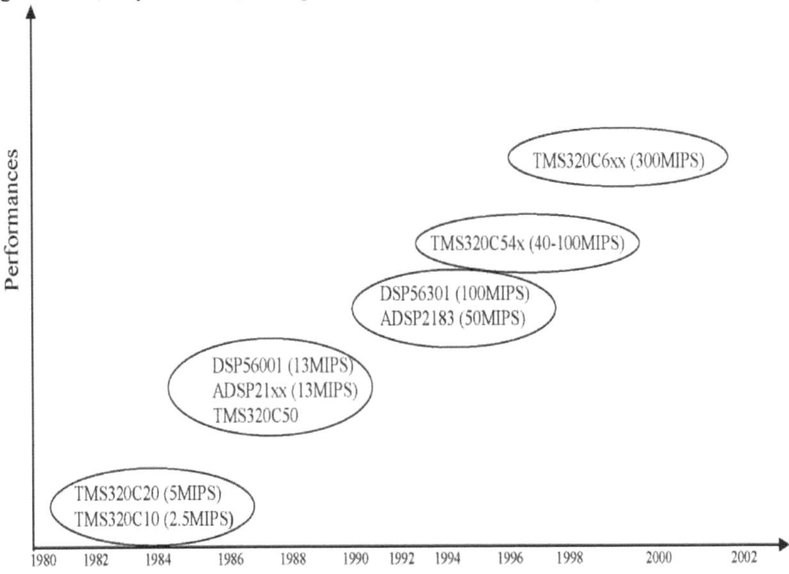


Fig1, The different families of DSP processors

# CHAPTER 2

## II. characteristics and architecture of conventional public service contracts

In principle, conventional microprocessors are not designed for specific applications. Their standard architecture is rather independent of the application being developed. DSPs, on the other hand, are optimised for calculating digital signal processing algorithms. This is a specific type of processing, where the calculation is at the origin of more complex arithmetic and logic operations (multiplication, accumulation, bit reverse addressing, etc.). By way of example, a conventional microprocessor (figure (2a)) requires several clock cycles to perform a given calculation. While this time is acceptable in the general case of current applications, it cannot be acceptable in the field of signal processing, where the notion of real time is very much in demand. DSPs are designed to optimise computation time, and therefore have a special architecture and specialised functions that enable them to execute algorithms more quickly. Data addressing modes are also a particular aspect of this type of architecture. Such processors have several address generation logic units working in parallel with other calculation units (figure (2b)) [5,6].

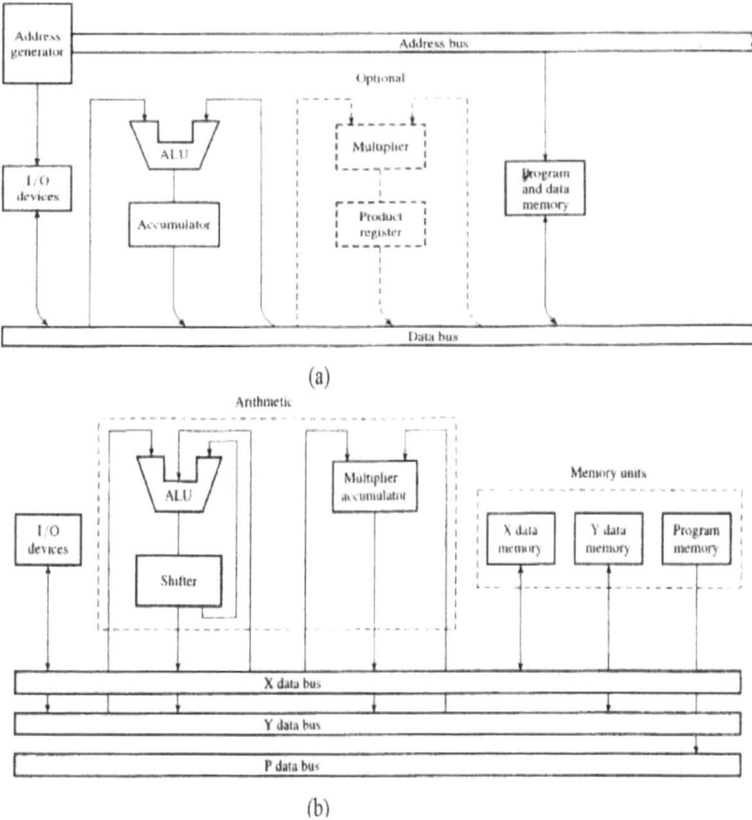



Fig.2, a). Classic processor architecture and b). DSP

The data path and memory architecture are the two main features that differentiate DSPs from

4

conventional processors.

## 11.1 Data path

### 11.1.1 Fixed / floating arithmetic

There are two distinct categories of DSPs: integer arithmetic (or fixed-point) processors, and floating-point arithmetic (or floating-point) processors. Integer processors are by far the most widely used, mainly because of their cost, see figure (3). We won't go into the mathematical principles of these two approaches here. We will simply recall that in fixed-point arithmetic, numbers are coded in 2's complement and can be of two types: integers or fractional (between -1 and 1). In floating-point arithmetic, a floating-point number uses three distinct fields: a sign bit, a mantissa and an exponent, figure (4) [2].

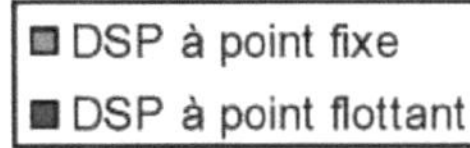


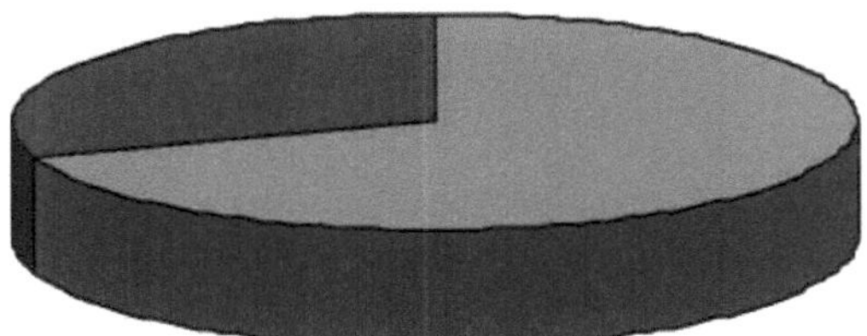

Fig. 3, Use of fixed point / floating point processors

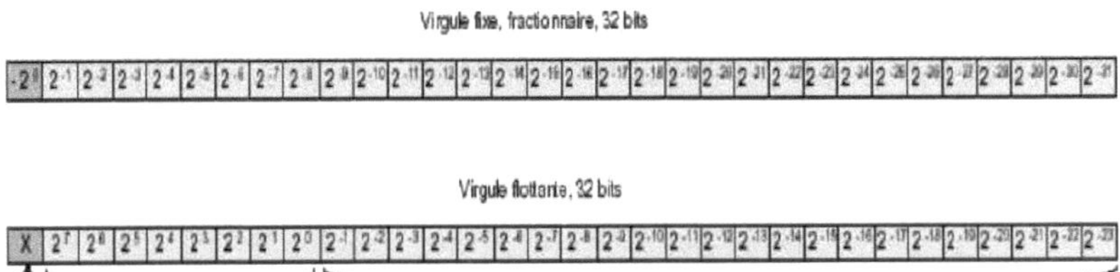


Fig. 4, Fixed and floating point number representation

## 11.1. 2 Data width

The native data width of a processor is the maximum width of data that can circulate on its buses. Most processors operate in 16/32bit, where these two numbers correspond to single/double precision data. Some processors also operate in 24/48 bits. The Arithmetic Logic Unit (ALU) and accumulators A and B of the processor in figure (5) are 56 bits wide for a native data width of 24 bits. In addition to the 48 bits encoding the result of the 24x24 multiplication, there are 8 extra bits called "guard bits". These prevent overflow errors when accumulating 48-bit values, allowing several MAC (Multiplication-Accumulation) operations to be chained together sequentially. This type of mechanism is widespread in most DSPs (16-bit processors use 40-bit accumulators) [7].

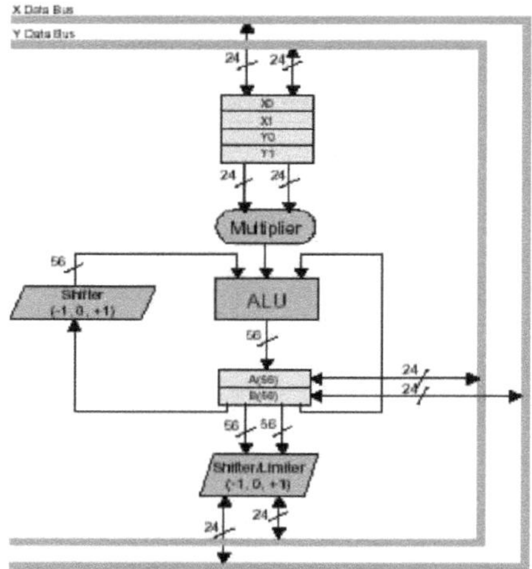


Fig. 5, $2^{ere}$ generation DSP architecture: the Motorola DSP5600x

### 11.2. 3 Functional units

Mainly dedicated to digital signal processing, DSPs have specialised functional units whose main task is the rapid execution of algorithms. Although there are many variations, these units can be classified into three main categories that are almost always found in all DSP processors:

- **Arithmetic and Logic Unit (ALU)**: This unit has the classic functions found in all processors, such as addition, subtraction, elementary logic operations, etc. It also incorporates more specific functions linked to precision management, such as overflow control and rounding functions, as well as advanced arithmetic operations such as absolute value, maximum/minimum of two values, iterative division calculation, etc.

- **Multiplier - Accumulator (MAC)**: This unit performs multiplication plus addition in a single instruction. Most DSPs perform these operations in a single machine cycle, which the most powerful microprocessors are generally unable to do. Some recent DSPs provide two or more multiplication-accumulation units, thus performing MAC operations in parallel. In addition, in order to carry out a series of multiply-accumulate operations without arithmetic overflow, DSPs generally provide additional "guard" bits in the accumulator. For example, Motorola's family of DSPs offers eight guard bits (DSP5600x case).

- **Shifters** and bit manipulation units (BMU): The purpose of this unit is to handle any possible shifts in the data. As with the other units, the data arrives via two temporary buffer registers and is then processed in the unit. It is designed to perform :
    - n-bit shifts to the left or right in a single clock cycle ;
    - word rotations ;
    - normalisation, which consists of converting numbers in fixed point format to floating point format. These operations require several clock cycles;
    - denormalisations (reverse operations).

Conventionally, when the DSP operates on N bits, the shift unit is divided between two registers, each comprising N bits. Shifting, whether to the right or to the left, assumes that the origin of this operation is fixed. The N-bit data is therefore loaded into either the least significant register or the most significant register. In some processors, it is possible to program more complex shifts, such as word rotation. The notion of high-order bits and low-order bits disappears, and it is then necessary to recover bits "on the left" and replace them "on the right" in the same reading order. Another important role for the unit is normalisation and denormalisation, which require shifts.

In standardisation, a number written in fixed-point format is converted to floating-point format in two clock cycles. The input data (in fixed point format) must be shifted to the left so that the most significant bit is the sign bit. The number of shift places required is taken into account and converted into a binary number by the exponent detector, then placed in a register (usually marked EXP). A priority encoder is used to perform this conversion. The first clock cycle performs the detection/conversion, and the second cycle performs the necessary offset.

In denormalisation, a number written in floating-point format is converted to fixed-point format in a single clock cycle. A word from the data bus is first loaded into the EXP register. This word determines the offset required to perform the conversion. The least significant bits of the mantissa are lost during the shift. By extension, denormalisation allows all kinds of word shifts, single or double precision, and even word rotations.

The operation on a floating point block (or BFP) consists of extracting the exponent on the largest number of data items available. Each exponent is associated with the entire block of data under consideration. When the first data item is passed, its exponent is detected and loaded into the Block Exponent register. The exponent of the next data item is evaluated, stored in the EXP register and compared with the Block Exponent register. The smaller of the two is then loaded into the Block Exponent register. The process continues until the end of the transmission of the block of data; the Block Exponent register then contains the most negative exponent possible. the operation on a floating point block is only a "scan". There is no real shift since the final value in the Block Exponent register can only be known at the end of the process. The advantage of this method is that it is possible to determine the best dynamics without sacrificing the accuracy of the mantissa.

## 1.1.4 4 Topology

In conventional microprocessors, data is stored internally in a register bank that is central to the architecture, often multi-port, and is used for both arithmetic and address calculations. In this type of architecture, each register in the data path has the same visibility vis-à-vis the functional units. From a programming point of view, this means that the registers are completely interchangeable. In this case we talk about a "homogeneous" set of registers [7].

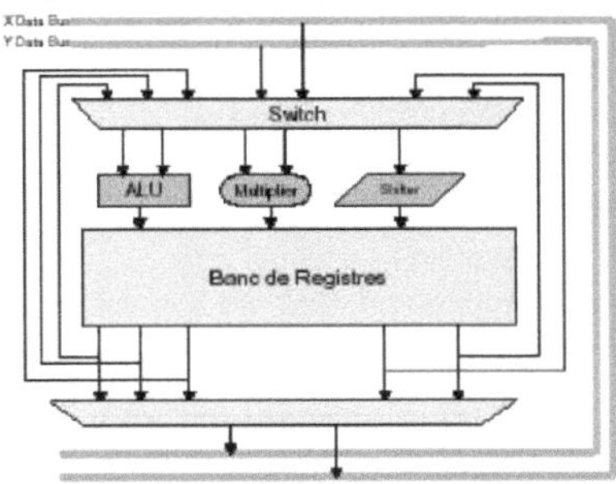


Fig. 6, Homogeneous data path

The strategy used in conventional DSPs is radically different. Some registers are dedicated to particular operators and can only be accessed as source operands or as the result of instructions

7

associated with the operator. Data located in the 'Accumulators' register bank can only be used as the source operand for a multiplication by first transferring this data to one of these registers.

Historically, the raison d'être for this type of architecture has been the quest for minimum hardware cost. DSPs have greater cost and power constraints than conventional microprocessors. The previous figure shows an architecture with a "homogeneous" set of registers, using a multi-port register bank (3 input ports, 5 output ports), making it possible to supply enough data to perform a multiplication, addition and shift in parallel. However, for the same number of registers, the register bank solution is much more expensive at all levels (in terms of performance, hardware and power consumption). What's more, as the bit width of the buses and registers is limited by the larger data sizes (here 56 bits at the output of the multiplier in the 56000 DSP example), the size of most of the elements in the data path increases from 24 bits to 56 bits, making this solution prohibitive overall in terms of cost and performance. Designers have therefore opted to use heterogeneous architectures that minimise hardware and maximise performance. The data path topologies chosen, although varying from processor to processor, all tend to optimise the execution of the most widely used DSP algorithms, FIR filters in particular.

The main disadvantage of this type of architecture is its lack of programming flexibility. As we pointed out earlier, data flow in this type of architecture is more complex than in the case of homogeneous structures because of the strong topological constraints: a given register is only connected to a given unit, a given result must go into a given register and not another. In addition, memory resources (internal registers) are reduced to a minimum, making the task of register allocation particularly tricky. This is why programming this type of DSP is very difficult and requires a great deal of attention and experience. This is also why compilers perform so poorly, forcing users to hand-code their applications directly in assembler.

## 11.2 Memory Architecture

### 11.2.1 Architecture Harvard

A major difference between conventional microprocessors and DSPs is the memory architecture, which defines how the processor core accesses data and instructions. Traditionally, microprocessors other than DSPs are based on a Von Neumann-type architecture, which uses a single bus for accessing data and instructions figure (7a). This solution, which is satisfactory for general-purpose applications, has its limitations in the case of signal processing applications. This is because most DSP algorithms require a higher memory bandwidth than the Von Neumann architecture can provide. For the FIR filter, for example, if you want to perform the equivalent of one tap per instruction cycle, the processor has to calculate a multiplication-accumulation and access the memory several times in the same instruction. Specifically, for each tap, the processor must :

- load multiplication instruction - accumulation ;
- read the appropriate sample from the delay line;
- read the coefficient of the current tap ;
- calculate the product coefficient x sample and add it to the previous result ;
- write the sample read in the next position on the delay line, in order to shift the samples in the line.

This represents a total of four memory accesses per cycle. In practice, various techniques make it possible to reduce this number to three or even two accesses per cycle, which in any case remains beyond the "1 access per cycle" limit of the Von Neumann architecture. A DSP processor with an arithmetic unit capable of performing a MAC operation and using an architecture of this type would be inefficient, since it would be unable to supply a sample and a coefficient per cycle, which are required to feed the MAC unit [8],

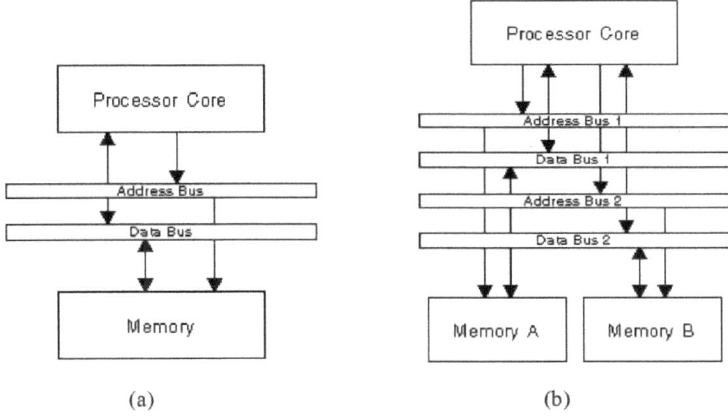


(a)                                                    (b)

Fig. 7, Memory architecture: a). Von Neumann and b). Harvard

This is why DSPs use a different type of memory architecture, known as the Harvard figure (7b). In this architecture, several address and data buses are used, allowing several simultaneous memory accesses per cycle. The classic Harvard architecture includes a set of Program (Address and Data) buses and a set of Data buses, the first being used to load the next instruction and the second to load a piece of data required by the current instruction, thus offering an effective throughput of one instruction and one piece of data per cycle. More advanced architectures, known as Enhanced Harvard, offer a further increase in the number of buses to provide higher bandwidths of 2, 4 or even 8 data items per cycle. This is the case with Motorola's DSP5600x (see Figure 8), which has three separate memory accesses and their associated buses: one for instructions and two for data (X and Y buses). This processor can therefore load one instruction and two data items per cycle.

**11.2.2 On Chip Memory**

Unlike conventional microprocessors, most DSPs do not use cache memory mechanisms, either for data or for instructions. The size of signal processing algorithms is generally small (a few tens of C lines), and the resulting code can fit into restricted memory spaces. As a result, the dominant philosophy for DSPs is to place the instruction memory directly on the circuit, generally in the form of fast SRAM-type memory responding in one cycle; this ensures a throughput of one instruction per cycle, the equivalent of an instruction cache. Space is also reserved on the circuit for blocks of data memory, again offering maximum throughput thanks to single-cycle accesses. Most DSP algorithms consume data in the form of streams: a sample is read, used for calculation and then the next sample is read and placed at the location of the previous one. The memory size required to store the data of a signal processing algorithm is therefore generally small, so that a small amount of data memory integrated directly into the circuit is sufficient to ensure optimum performance. Data is often accessed via dual-port memories, which are more flexible in use but more expensive. The solutions adopted often combine the two types of memory [7]. It should be noted, however, that the absence of caches also avoids the problems of unpredictability of execution time due to the dynamic control mechanisms characteristic of superscalar architectures. However, we will see in the next section that some of the most recent processors have adopted cache mechanisms in order to take account of the increasing complexity of applications. Since memory blocks integrated directly on the circuit are very expensive in terms of silicon, DSPs always incorporate interfaces for external memories, which make it possible to increase the system's memory capacity. Access to data or instructions in external memory is obviously much slower, which is why the efficient implementation of an algorithm requires rigorous partitioning of data and instructions in the different memory banks, with the most frequently used parts of the code and data being located on the circuit as far as possible, while the rest of the code and data that is not used very much can be stored externally. External memory is often accessed via a single external bus, with the various internal buses being multiplexed.

**11.3 Address calculation units**

The intensive calculation loops of DSP algorithms make very frequent accesses to memory to load source data or store results. The FIR filter, for example, requires a sample and a coefficient to be loaded for each tap calculation in order to perform multiplication - accumulation. The pointers must then be incremented to point to the next sample and coefficient. After calculating the last tap, the pointers must be reset to point to the first coefficient (cl) and the new starting sample before starting the calculation of the next "y". To speed up execution, most DSPs are capable of calculating the MAC operation and modifying the pointers in a single machine cycle. This means that they have to perform 3 arithmetic operations in one cycle: the MAC operation and the two pointer increments. As the data path hardware is already used by the MAC operation, address calculations are handled by specialised units: Address Generation Units (AGUs). These units work in parallel with the data path, relieving it of all the operations involved in addressing the data. Figure (8) illustrates the architecture of an AGU used in the ADSP21xx from Analog Devices. The address pointers are stored in the I register bank and the arithmetic operations on the pointers (additions and subtractions) are calculated by the ADD unit [9],

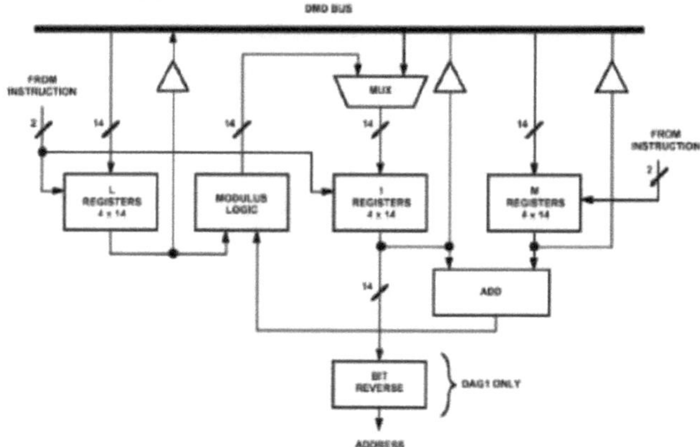


Fig. 8, ADSP21xx address calculation unit

DSP AGUs manage the classic addressing modes used in most processors: immediate addressing, direct addressing, indirect addressing (by register). Most also offer indexed indirect addressing. But above all, they have additional functions specific to DSP calculations:
- auto incrementing or decrementing of pointers ;
- modulo addressing ;
- bit - reverse addressing ;

The assembly code for the execution of the FIR filter on DSP in Figure (9a) shows the instruction "xO = *ptrO++", which represents the loading into the xO register of the value pointed to by ptrO, followed by the automatic incrementation of the pointer value. Most DSP algorithms work on data streams and access the data serially, which makes this addressing particularly practical, as in the case of the FIR filter, where the pointer to the coefficients must move by one unit with each cycle (figure9b). At the end of the calculation of an output result, the pointer to the coefficients must be initialised at the start of the table. Modulo addressing avoids the extra cost associated with reset instructions: it is the hardware which detects that the pointer is at the end of the memory area allocated to the coefficients, and the next increment instruction will bring the pointer back to the start address of the coefficient table. In the case of the ADSP21xx, the operation is as follows: an address register I is assigned to the current pointer, the length N of the array is stored in a register of length L, and the increment step is stored in a register M (it is possible to have increment steps other than 1). For post-increment addressing, the AGU calculates the sum of the value of the current pointer with that of the increment and compares it with the end-of-array address (this is the role of the modulo block); in the event of an overflow, the modulo block returns the address modulo the length of the filter, causing the pointer to be returned to the beginning of the array, without any software overhead.

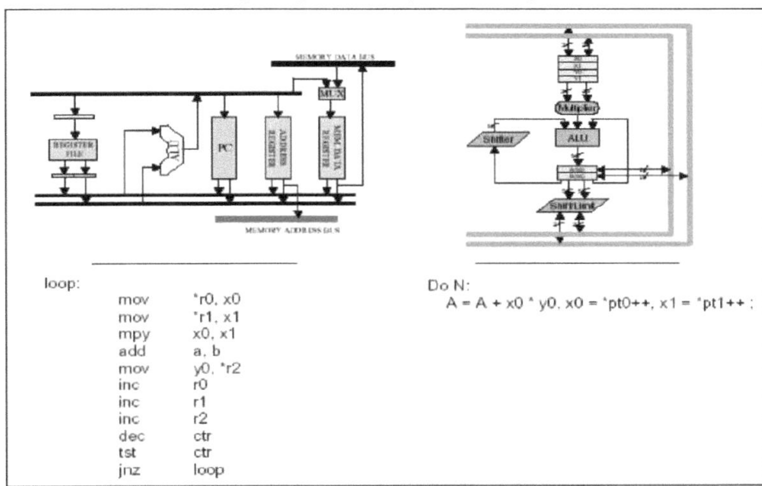


Fig.9a, FIR filter on a RISC-type architecture and a DSP.

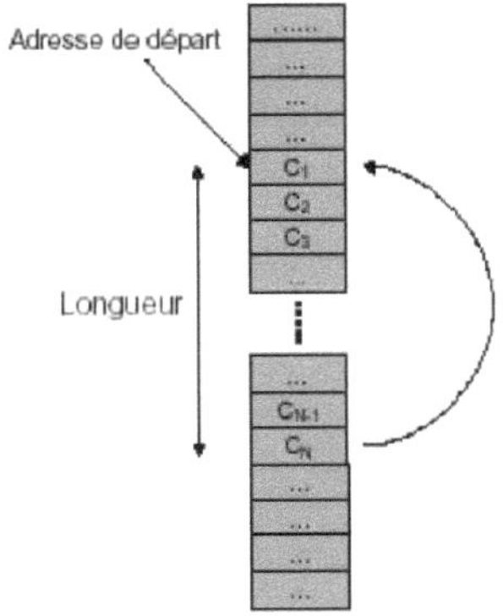


Fig.9b, Principle of modulo addressing

11

Bit Reverse addressing is a special type of addressing used in the Fast Fourier Transform (FFT) algorithm, which manipulates samples in an order other than sequential. The ADSP21xx's Bit Reverse wired block transforms sequential addressing into Bit Reverse addressing, eliminating the need for software address calculations. Since FFT is used extensively in signal processing, this feature is found in all AGUs.

## 11.4 Control unit

The DSP control unit, which supervises the flow of instructions to be executed, obviously manages the basic processor control modes: sequential execution, conditional jumps and branches, subroutine calls and interrupts. Since DSP algorithms make intensive use of software loops (most often for loops), DSP processors have hardware mechanisms to speed up the execution of these loops. On a conventional processor, a software loop is coded using a conditional jump instruction preceded by a decrement and a loop counter test (the FIR filter loop (see fig.9a). On a DSP processor, all these operations are transparent: all you have to do is indicate the start and end of the loop, initialise a loop counter in software, and the operations of changing the counter value and jumping to the start of the loop will be carried out by the hardware without any loss of machine cycles, hence the name "zero-overhead loop". Some processors only allow single instruction loops, but all modern processors allow loops longer than 1 (multiple instruction loops). They also allow several levels of nested loops, the number of which is variable and limited by the number of internal registers storing the characteristics of the different loops. To reduce the number of cycles due to sub-programme calls and returns, processors sometimes have a hardware stack which stores the return addresses of sub-programmes or interrupts, enabling rapid returns in one cycle (no access to the software stack).

Some processors also use shadow registers to save and automatically restore the context; this makes it possible to lighten and speed up interrupt processing routines, which only have to save part of the context. For critical interrupt routines that are called regularly (such as those generated by a serial interface-type data acquisition peripheral), the performance gain can be very significant. Finally, DSPs often integrate additional logic into the circuit to facilitate real-time debugging of applications (for some DSP systems, debugging can only be done in real time). The technique used is generally based on a JTAG interface that allows registers and circuit memory to be read and modified, and breakpoints to be set, all without disturbing the dynamic behaviour of the application running on the processor.

## 11.5 Instruction set

Conventional microprocessors use RISC (*Reduced Instruction Set Computer*) instruction sets with general functionality, in which an instruction corresponds to a single operation. In contrast, DSP processors have an instruction set containing specialised instructions: complex arithmetic operations, specific addressing modes, and so on. What's more, these instructions encode several operations in parallel, such as the FIR filter instruction in the (see fig.9b). The architecture of conventional DSPs has a small number of execution units (ALUs, multipliers and AGUs) and uses a heterogeneous set of registers (the output of a functional unit is sometimes connected to only one register), so the number of bits needed to encode an operation is less than that required by a more general, homogeneous architecture. The term "complex encoding" is used to express the fact that several operations are encoded in a reduced number of bits. This type of instruction set leads to very restrictive programming because not all combinations of elementary operations necessarily correspond to an existing instruction.

## 11.6 Specialised peripherals

With the exception of processor cores consisting solely of computing units and possibly memory blocks, on-board DSP processors incorporate specialised peripherals and advanced

input/output interfaces on the silicon, in order to reduce the overall cost of the system. The most commonly used peripherals include :
- serial ports: these are usually synchronous and are used for communication between the processor and external peripherals (e.g. Digital to Analogue Converters (DACs) or Digital to Analogue Converters (DACs)), or between processors; - parallel ports;
- Timers: used to generate periodic interrupts. They use a clock divider, sometimes connected externally, to act as a variable frequency generator;
- host ports: specialised ports for connection to a conventional microprocessor or another DSP. They are used to exchange data between processors or to control the operation of the DSP;
- communication ports: special parallel ports for multiprocessor communication. They differ from parallel ports and host ports in two ways: they do not allow processors to be controlled, and the processors connected must all be of the same type;
- bit" input/output ports: 1 bit wide, they can be configured as input or output; generally used for control, they can also be used for data transfer;
- interrupt lines: used by external peripherals to generate interrupts to the processor (two types exist: on edge or on level);
- converters (ADC and DAC): generally with 16-bit resolution, they are used in certain DSPs used in speech processing (mobile telephony);
- DMA controllers: the use of DMA channels relieves the processor of the management of memory-to-memory or memory-to-device transfers. Once the DMA controller has been configured by specifying the address and size of the area to be transferred, it takes control of the bus and accesses the peripherals and memory itself. Depending on the number of memory banks and internal buses, the processor is either frozen for the duration of the transfer or can continue to work normally if there is sufficient memory bandwidth remaining. Some DMA controllers are also capable of managing several DMA channels in parallel.

## III. NEW DSP ARCHITECTURES

Until the mid-1990s, most DSPs were based on a conventional architecture with specialised functional units, a complex instruction set and instruction cycle execution.

The advances in performance of these processors owe more to changes in manufacturing processes than to any real architectural upheaval.

Today, the offer is tending to diversify with new architectures borrowing from general microprocessors: VLIW (very long instruction word), superscalar, and SIMD (Single Instruction on Multiple Data).

The reason for this upheaval is that conventional DSPs have reached their limits and cannot meet the demands of modern applications, which are increasingly demanding in terms of computing power.

Indeed, the complexity of DSP applications continues to grow rapidly and the role of compilers, very often neglected by DSP designers and users, is becoming essential to maintain reasonable development times.

At the same time, the main constraint for processor designers is now power consumption. Simply increasing processor frequency to improve performance (the main factor driving DSP performance growth in recent years) leads to over-consumption, which is no longer tolerable from the point of view of embedded systems.

The current trend is therefore to reduce operating frequencies in favour of greater hardware parallelism, to reduce power consumption while maintaining performance levels.

Silicon surface area is a concern that now takes second place to power consumption. Texas Instruments was the first manufacturer to introduce a VLIW architecture in a DSP processor with the TMS320C6x, reflecting an evolution towards the use of more advanced architectural techniques than the simple architecture based on the structure of digital filters.

Certain techniques, such as superscalar, are widely known and used in the field of general microprocessors, where they have proved their effectiveness.

Conversely, the VLIW and SIMD architectures, although old from a theoretical point of view, have been less successful but seem promising for integration into signal processing systems.

The basic idea behind these techniques is to increase performance by taking advantage of the parallelism of instructions and data present in applications; in fact, the characteristics of conventional DSP architectures (one instruction per cycle, low number of functional units, complex instruction set) are unsuitable for the optimal exploitation of parallelism.

### III .1 Conventional "extended" PSAs

An initial solution to increasing the performance of DSPs was to improve the conventional architecture so that it could exploit more parallelism (Figure 10). Increasing integration densities, offering ever more transistors, allowed designers to add hardware to their architecture: - adding functional units ;

- larger memory bandwidth ;
- limited or extended SIMD capabilities ;
- specialised data path for a particular field of application ;
- hardware coprocessors.

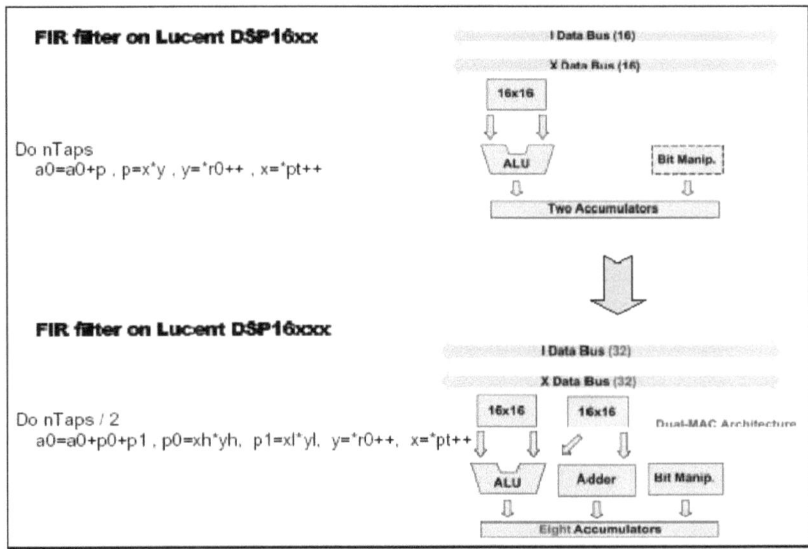


Fig. 10, Evolution of a conventional architecture: from 16xx to 16xxx

The Lucent DSP16xxx processor [10], successor to the 16xx, is a good example of the extension of a classic architecture. The hardware integrated in the data path has been increased: addition of an ALU and a multiplier, doubling the width of the memory buses, and increasing the number of registers. As the processor is aimed at GSM-type Telecom applications, the data path has been specialised in various ways:
- integration of limited-capacity shifters and saturators at the output of the functional units to automatically manage scaling operations.
- Specialised ALU capable of 16/32-bit SIMD operations and the ACS (Add / Compare / Select) operation used for Viterbi decoding.
- Trace back encoder coprocessor for accelerating Viterbi decoding. This choice of extending an existing architecture and instruction set enables a significant increase in performance while retaining most of the advantages of conventional DSPs: low power consumption, low hardware cost and good code density. Indeed, the specialisation of the data path (heterogeneous register set, non-regular topology, specialised units) tends to reduce hardware and power consumption to a minimum, to the detriment of the genericity of the architecture. The instruction set encodes even more elementary operations per instruction on a limited number of bits, generally 16, giving excellent results in terms of code density. Above all, this code remains compatible with that of older-generation processors, avoiding the need for users to reprogram an application that has already been implemented on a DSP. Given that most DSP applications are programmed manually in assembler for performance reasons, the possibility of reusing part of the code (and possibly optimising it for the new architecture) and the fact that the programming environment and instruction set are very similar, considerably reduces development time.
The disadvantages of the extended architecture are the same as those of the traditional architecture, and even greater: the increased complexity of the instruction set makes the work of compilers even more difficult, while manual programming in assembler, already complex, becomes even less obvious

because of the management of parallelism and the many control bits configuring the data path. Finally, the prospects for developing this type of architecture to exploit even more parallelism are very limited; adding two additional multipliers to a DSPlôxxx-type architecture would result in an architecture and instruction set that are abominably complex, and therefore unusable. Other examples of extended DSPs are the ADSP-2ll6x processor from Analog Devices, and the PalmDSPCore from DSP Group [ll].

### 3.2 .2 VLIW and superscalar DSPs

The arrival in 1997 of Texas Instruments' TMS320C62, based on a VLIW architecture, represented a minor revolution in the world of DSPs [12]; it was in fact the first time that a DSP had freed itself from the traditional architecture and instruction set of conventional processors and ventured down a new path. The reasons for abandoning the conventional architecture can be explained by the designers' desire to go further in exploiting instruction parallelism, but also to offer a more flexible, less complex architecture that could be more easily exploited by a compiler (figurell).

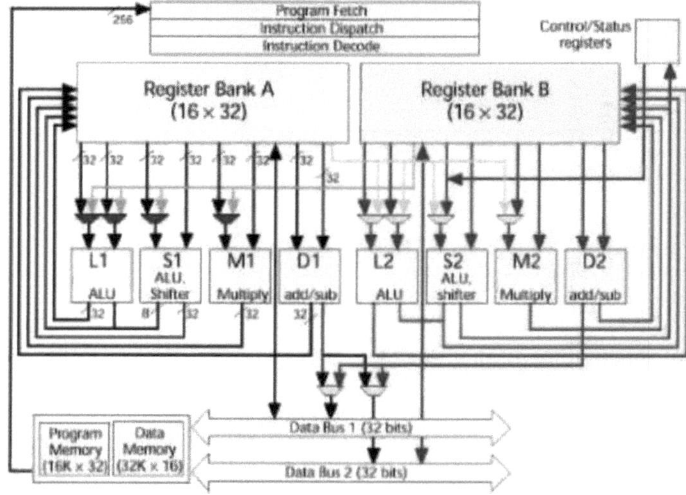


Fig. 11, An example of VLIW architecture: the TMS320C62x

### III.2 THE      .1 Architecture

The VLIW concept is based on a regular, homogeneous architecture made up of functional units arranged in parallel and multi-port register banks for storing temporary calculations. The TMS320C62, shown in Figure (11), is made up of eight functional units (4 ALUs, 2 multipliers, 2 adders-subtractors mainly responsible for address calculations), distributed in two "clusters" of 4 units, each with a multi-port register bank. The bandwidth to the data memory is high (2 * 32 bits per cycle) so as to supply the many functional units efficiently. There are many architecture variants, particularly in terms of the choice and number of functional units, the way the data path is structured into one or more clusters, the number of register banks and the memory bandwidth. In all cases, the philosophy remains the same:
- regular data path, made up of functional units working in parallel ;
- homogeneous set of registers made up of multiport register banks ;
- high memory bandwidth.

Unlike conventional architectures, VLIW processors do not have special registers associated with a single operator or 'vertical' chaining of certain functional units. In fact, these characteristics stem from the specialisation of architectures for well-defined application domains, whereas VLIW architectures are by nature much more general. The data path of superscalar processors is based on the same principles as those of VLIWs. It is mainly at the instruction set level that these processors differ [12].

### 3.2.1 .2 Instruction set

We saw earlier that conventional instruction sets encode numerous elementary operations (equivalent to a RISC (Reduced Instruction Set Computer) processor instruction) into an instruction of reduced width in order to offer instructions that are both high-performance and take up little memory space. The low number of bits required to encode an instruction is due to a complex encoding strategy that only allows certain combinations of elementary operations, the chosen combinations being selected on the basis of the study of the most classic DSP algorithms. If a combination of operations is 'out of the ordinary', there may be no equivalent in the instruction set and it will have to be performed sequentially using simpler but less efficient instructions. The DSP16xx, for example, does not allow the simultaneous execution of an exponent calculation and two multiplications, because this instruction is not used, or is only rarely used, in most conventional DSP algorithms. It is the instruction decoder that extracts the various commands for the functional units from the instruction word. Because of encoding constraints, it is rare that all the units can be called by a single instruction (Figure 12).

The strategy of VLIW processors as superscalar, on the other hand, is based on a much simpler set of elementary instructions equivalent to a RISC-type instruction set [6]. To increase performance, these processors allow several of these instructions to be executed in parallel in the same machine cycle. The set of permissible elementary instruction combinations is much larger than in the case of the highly encoded instruction sets of conventional processors. Instruction encoding in a VLIW processor is simple: the elementary instructions, each of which corresponds to the activation of one of the processor's functional units, are concatenated to form a super-instruction that will control the entire data path. This super-instruction is then processed by the dispatch unit, which isolates the fields associated with the units and propagates the commands to these units. The large width of the instruction word means that all the functional units can be activated simultaneously to take advantage of maximum parallelism [13].

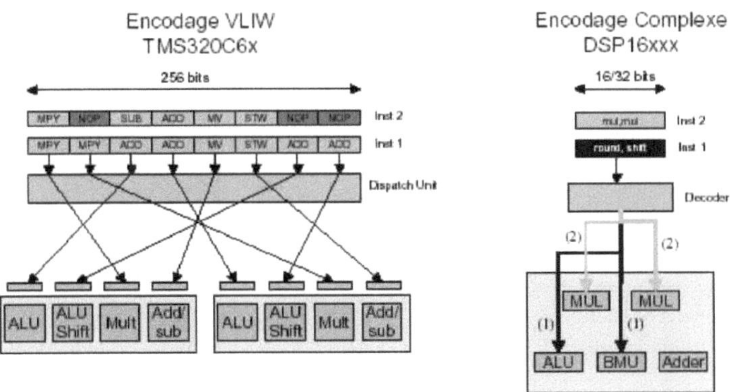


Fig.12, Instruction encoding

### 3.2.2 .3 Differences between VLIW and superscalar

The main difference between superscalar and VLIW processors lies in the representation and management of instruction parallelism.

In a VLIW architecture, parallelism is analysed statically before execution by the programmer or compiler, then translated into a sequence of super-instructions arranged in order in program memory.

Because of the size of the super-instructions, the width of the program memory and associated buses is much greater than for conventional processors (256 bits for the TMS320C6x). When executing a program, the VLIW processor loads the super-instructions one by one and executes them in the same order as a RISC processor.

In superscalar processors, parallelism is analysed dynamically by the processor itself. The programme is stored in memory in the form of a sequence of elementary RISC-type instructions (sequence generated by the programmer or compiler). During execution, the instructions are loaded by the processor into an instruction buffer.

Complex hardware mechanisms analyse the constraints between the different instructions to determine which can be executed in parallel.

These constraints take into account both the data and control dependencies between the different instructions and the occupation of processor resources. In this phase, the instructions can be reordered to minimise the constraints and thus offer more usable parallelism. A certain number of instructions are then chosen, grouped together in the form of an execution package and executed in the same way as a VLIW instruction (Figure 13).

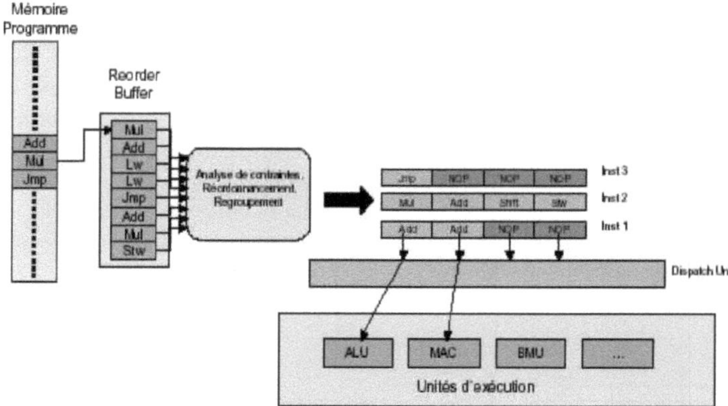


Fig 13, Execution on a superscalar processor

### III .3 SIMD features

Unlike the conventional terms, VLIW or superscalar, which characterise a type of processor through its architecture and instruction set, the term SIMD (for Single Instruction Multiple Data) simply expresses a processor's ability to exploit data parallelism. This functionality was already available on some conventional DSPs in a fairly basic form, that of the so-called 'Split-ALU' functional units, which enabled a 32-bit ALU to perform a choice of 2 16-bit operations or one 32-bit operation.

This type of low-cost capability in terms of hardware complexity can be found in most DSPs dedicated to the telephony sector, where data is coded on 16 and 32 bits (Figure 14).

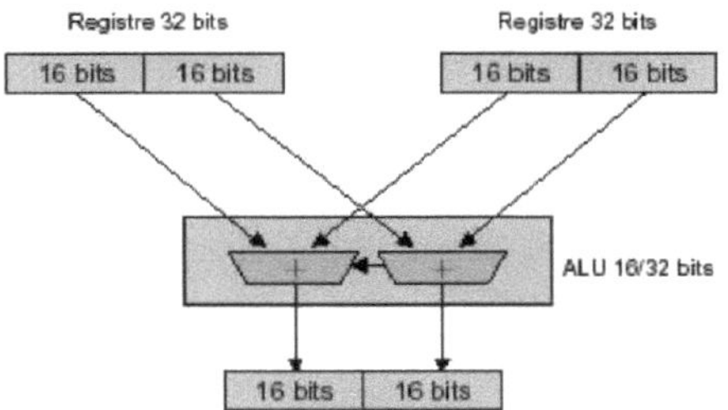


Fig.14, Split-ALU 16/32 bits

This type of SIMD processing, which consists of dividing the width of the operand data and applying the same processing to each "piece" obtained, is known as "subword parallelism" (SWP). This is the most basic form of data parallelism, with a granularity of just a few instructions (2 in the case of a 16/32-bit SIMD instruction). The other form of data parallelism is based on the parallelism of larger loops and has so far only been implemented in one DSP processor. This is Analog Device's TigerSHARC [14], whose instruction set includes so-called "hierarchical" SIMD instructions which handle both forms of data parallelism: SWP at the level of functional units, and a higher level of granularity "data path" parallelism (Figure 15).

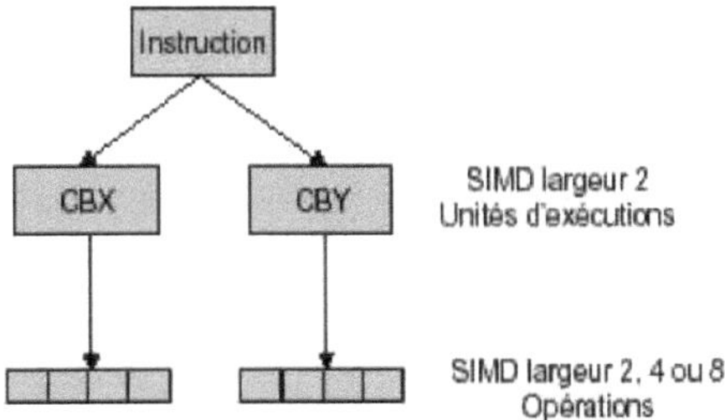


Fig.15, Hierarchical SIMD instructions for TigerSHARC

## III.4 Hybrid RISC / DSP architectures

Most specialist mobile phone circuits often incorporate both a microcontroller (or embedded microprocessor) and a DSP processor. The former is responsible for all control tasks (user interface, system synchronisation), while the latter is used to run DSP applications that require high computing power and therefore a specialised architecture. This dual-processor approach has a number of disadvantages: the need to develop code for two different instruction sets, the cost of using two processors (footprint, power consumption, price), the difficulty of interfacing, and so on. Some manufacturers have therefore turned their attention to solutions that allow the tasks assigned to the microprocessor and the DSP processor to be carried out with a single circuit. From an architectural point of view, the problem comes down to making control and DSP processing capabilities coexist in an efficient and inexpensive way. Several solutions have been proposed:
- RISC microcontroller + DSP coprocessor. For example, the Massana FILU-200 [15]. The microcontroller carries out most of the processing and delegates the DSP-oriented calculations to the specialised coprocessor. The two cores can work in parallel to improve performance. The main disadvantage is the potential complexity of programming: handling two instruction sets and

effectively managing communications between the cores. The complexity of the interface between the two cores can make RISC-coprocessor communications very costly in terms of execution time.

- Microcontroller with DSP capabilities. For example, the SH-DSP from Hitachi. This circuit is an extension of the SH-2 microcontroller, to which a data path and specialised DSP instructions have been added. A single instruction set controls both data paths, which simplifies programming and allows the microcontroller code to be reused, but does not allow both data paths to operate simultaneously, which limits performance.

- DSP processor with control capabilities. For example, the TMS320C27x. This solution has the same shortcomings and advantages as the microcontroller + DSP capabilities option, except that it favours the DSP side over control capabilities.

- Processor with mixed DSP/Control architecture. An example is Infineon's TriCore [16]. This processor is based on a superscalar architecture integrating specialised functional units for DSP calculations and developed control capabilities. As the proposed architecture is not a modification of an existing architecture, it is more homogeneous and therefore potentially easier to use. On the other hand, it no longer allows existing code to be reused. Whatever the solution adopted, RISC/DSP hybrids are in fact compromises between resolutely DSP-oriented architectures and others that are definitely specialised in control. As a result, their raw performance cannot rival that of 'high-performance' DSP processors, and is more in the region of that of entry-level DSPs. They are, however, of interest for systems requiring dual control/DSP skills, provided that the performance required is not too high.

# IV. OTHER COMPUTING ARCHITECTURES

Signal processing algorithms can be handled by several types of architecture. These can range from simple low-end microprocessors to highly specialised ASICs. However, the implementation of most algorithms involves constraints that are difficult to meet by non-specialised architectures. The complexity of DSPs varies according to the application. The first question the designer needs to ask is which systems can support the load required by the application. He must then take into account the other constraints in the specifications (cost, power, development time, etc.) in order to select the appropriate system. The wrong choice can have serious consequences for the technical or economic viability of the circuit [17].

## V.   .1 Microcontroller or low-end microprocessor

This type of circuit offers poor performance, mainly for the following reasons:
- Poor arithmetic operators (especially for multiplication), no precision management (overflow, rounding).
- Significant delay for address calculation or loop management.
- Limited memory bandwidth.

## IV.2 Latest-generation microprocessors

These are currently available general-purpose processors (*Pentiums*) whose highly advanced architectures can deliver very good performance in numerical computation, due to the following characteristics:
- Very high clock frequencies (100 MHz);
- High-performance arithmetic operations (execution in one cycle) ;
- Good memory bandwidth (use of cache memory) ;
- Superscalar architecture, use of dynamic execution techniques (branch prediction, instruction reordering, etc.).

The main feature of the Pentium compared with its predecessors is the use of an internal cache memory called L1 of 8 KB for programmes and 8 KB for data, directly installed in the processor.

## IV.2.1 Pentium II

The PENTIUM II processor has its own unique features. Firstly, the L2 cache memory is no longer located on the motherboard, but directly on the processor casing, and works at half the processor's internal frequency. First of all, there are the Pentium IIs with frequencies of 233, 266 and 300 MHz. The chipset is the 440LX (originally 440FX). It manages Dimm memory at 66 MHz, the AGP screen bus, the Ultra-ATA hard disk interface at 33 MB/s and the external USB bus. The size of a Pentium transistor is 0.25 microns.

As INTEL decided to abandon the PENTIUM MMX and Pentium II prices were too high, INTEL released the CELERON for low-end machines. Initially, it did not include cache memory. In recent years, it has been 128K, but is clocked at the same speed as the INTEL processor, which has released a 440EX (fin98) and 440ZX chipset that manages fewer slots for the PCI (3) and ISA (1) buses. The Pentiums that followed were at 333 (released in March 1998), then 350, 400, 450, 500 Mhz, ... These use the 440BX chipset, which manages memory at 100 MHz.

## IV.2.2 Pentium IIIs

Released in early 1999, the Pentium III is identical to the Pentium II, but incorporates additional multimedia instructions called SSE. With the release of the CAMINO 820i chipset at the end of 1999, the PENTIUM III coppermine are engraved in 0.18 microns, use Dimm 133 memory (via an interface) and DRDRAM at 300 (PC600, 1.6GB/s) and 400 Mhz (PC800). Figure 1.16 shows the architecture of the PENTIUM III.

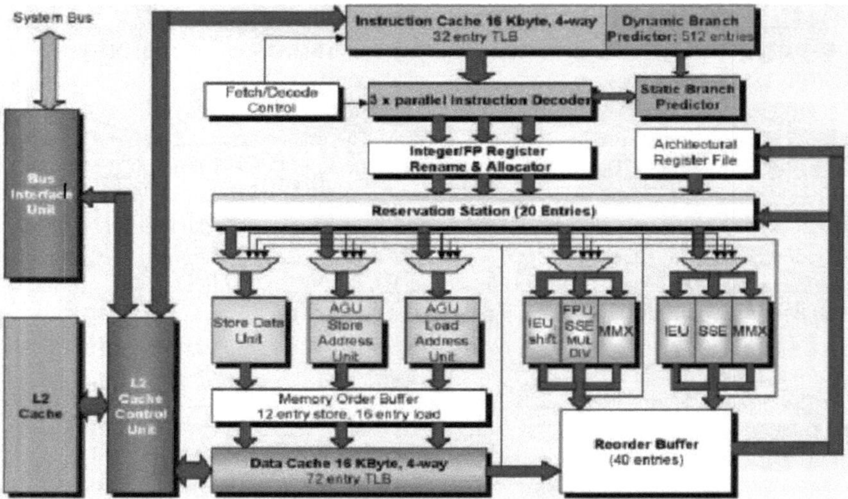

Fig.16, PENTIUM III architecture

For comparison purposes, Table .1 shows the main technological characteristics that differentiate the two processors, PII and PIII.

|  | **PENTIUM II** | **PENTIUM III** |
|---|---|---|
| 3D instructions | No | Yes (SSE) |
| Frequencies (Mhz) | 300/333/350/400/450 | 450/500/550/600 and more |
| L1 cache (KB) Prog + data | 16k+16k | 16k+16k |
| L2 cache speed | $\frac{1}{2}$ CPU speed | $\frac{1}{2}$ CPU speed |
| Quantity of cover L2 | 512K | 512K |
| Manufacturing technology (in pm) | 0,25 | 0,25 |
| CPU size (mm )$^2$ | 203 | 203 |
| Number of transistors (millions) | 7,5 | 7,5 |

Table 1, Characteristics of Pentium II and III.

Analysing the performance of different configurations is difficult. According to test benchmarks, for example, a K6-III running at 400 Mhz is 10% slower than a Pentium III at 400 Mhz. On the other hand, a 450 Mhz K6-III outperforms a 500 Mhz Pentium III by 10%.

### IV.2.3 The Pentium IV

While the Pentiums II and III use the same internal architecture, INTEL is releasing a brand new processor called the PENTIUM IV based on an architecture called NetBurst (fig.17). The main features of the Pentium IV are as follows:
- The architecture remains 32-bit.
- New SSE2 instructions used by DirectX 8.0, 144 instructions.

22

- Pipeline with 20 levels compared with 10 for Pentium IIIs, which has consequences in the event of incorrect connection prediction.
- Modified calculation unit (2 ALUs running at twice the processor's internal speed and a floating point PGU).
- L2 cache memory remains at 256K (increased to 512K in early 2002) but has been improved. Here again, the bandwidth has been increased from 14.9 GB/s for a PIII running at 1 GHz to 41.7 GB/s for a P4 running at 1.4 GHz.
- The L1 cache now contains only an 8K data cache and an Instruction Trace Cache, which stores instructions after they have been decoded in RISC. This program cache can hold up to 12,000 instructions.
- The (external) bus frequency is 200 MHz, but will be increased to 400 MHz in early 2002.
- The engraving is0.18p, rising to0.15p in early 2002.

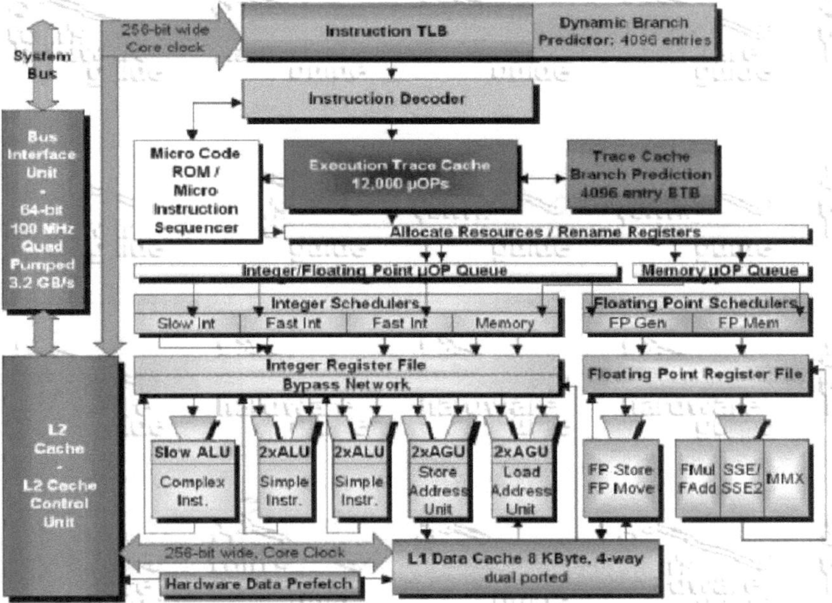



Fig.17, PENTIUM IV architecture

However, implementing signal processing algorithms on this type of processor is made difficult by the lack of signal processing-oriented development tools. In addition, dynamic execution results in non-deterministic temporal behaviour, making it difficult to develop a real-time application. Finally, the power consumption and cost/performance ratio of this type of processor are higher than those of specialised processors, particularly in the case of fixed-point applications. In recent years, there has been a real desire on the part of major processor manufacturers to break into the field of signal processing, and ultimately to compete with DSPs in a market that is currently reserved for them.

**IV.3 ASICs**

This type of circuit is generally used when the desired processing requires very high performance, which general architectures cannot provide. In this case, the architecture implemented is optimised according to the processing algorithm. This solution, the absolute best performer, suffers from a number of major drawbacks: a simple change in the application specification can mean almost complete redesign of the circuit, and the cost, development time and lack of flexibility are the reasons for this [17].

## IV.4 FPGAs

FPGA (*Field Programmable Gate Arrays*) manufacturers are now offering signal processing macrocells that can be directly implemented on their circuits. The size of these cells can often be parameterised, and they range from basic blocks (adders, multipliers, shifters, etc.) to more complex functions (floating-point operators, FIR filters). Inexpensive and quick to implement. The configurability aspect obviously comes at a cost: the silicon surface area and power consumption of an FPGA solution are much higher than those of its ASIC equivalent. Programmable interconnects, for example, use most of the surface area of an FPGA, whereas an ASIC will use a limited number of surface-optimised connections. What's more, implementing an algorithm on an FPGA never uses 100% of the circuit's capacity, leaving some of the hardware unused (but still consuming power!). Finally, FPGAs suffer from the same limitations as ASICs: they are primarily effective for medium-complexity, data-flow-oriented applications. To overcome this limitation, one solution is to add a processor core to the FPGA circuit to manage the 'control' parts of the application. However, this solution cannot currently deliver the performance of DSPs or ASICs [17].

## IV.5 Comparison table

Table 2 compares the hardware solutions available for implementing a digital signal processing system. Each type of solution has its advantages and disadvantages.

| Technology | Performance | Surface | Consumption | Flexibility | Design time |
|---|---|---|---|---|---|
| ASIC | Excellent | Low | Low | None | Very long |
| FPGA | Very good | Grande | Grande | Good | Short |
| DSP processor | Good | Average | Average | Very good | Average |
| Microprocessor / Microcontrollers | Average / Wrong | Average | Average | Very good | Average |

Table.2, Hardware solutions for different architectures

In terms of raw computing performance, the advantage goes to parallel hardware solutions such as ASICs and FPGAs. ASIC and FPGA solutions are not only the fastest, but also the ones with the lowest *area * time* cost, due to their excellent *computing density*. Computing density refers to the number of operations that can be performed per unit area. From this point of view, the weakness of programmable processors is due to the fact that they use large-capacity memories to store the software code. These memories are expensive in terms of silicon surface area, often more expensive than the processor itself. ASICs, on the other hand, have a hard-wired static control that takes up very little space. As for FPGAs, reconfigurability is ensured by the configuration bits of the CLBs (*configurable logic blocks*) and interconnections, which form a single very large instruction for the entire circuit. The better use of hardware in wired solutions is also reflected in power consumption. This parameter is difficult to estimate for a complete circuit, but is very closely linked to the silicon surface area. FPGAs and ASICs, which are denser in terms of surface area and have a much higher resource utilisation rate than processors, therefore have much better performance/consumption ratios, as shown in Figure 18.

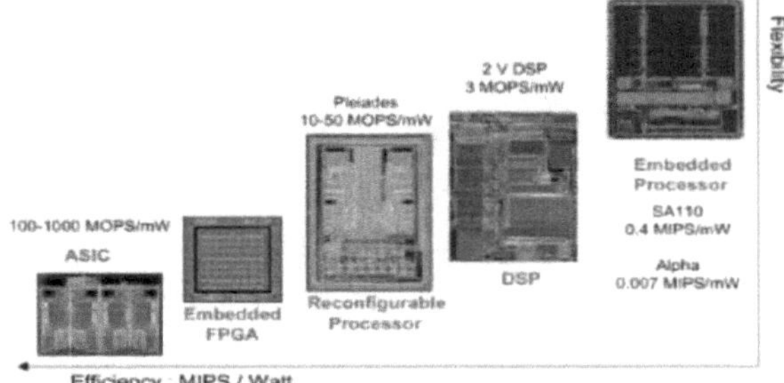


Fig.18, Performance, power consumption and flexibility of different hardware architectures

However, the use of specialised programmable processors instead of purely wired solutions is becoming increasingly common, particularly in the field of circuits dedicated to DSP applications. The increasing complexity of applications is one explanation. The final argument in favour of programmable systems is their flexibility, which makes it possible to take account of changes in standards or to correct design errors. In the case of a programmable system, upgrading simply involves updating the on-board software. In the case of a wired system, the behaviour of the architecture is fixed and cannot be modified. Taking account of a change in standard therefore means changing the whole system and redesigning a circuit that implements the new standard. Given the complexity and design time of an ASIC, the cost of upgrading is often prohibitive. As a result, the general trend in embedded system design is towards the use of mixed hardware/software solutions, in which the software component is increasingly predominant.

# V. DSP SOFTWARE DEVELOPMENT TOOLS (CODE COMPOSER)

As the use of DSPs is relatively sophisticated, it is important to have high-performance development tools that can minimise the time it takes to bring a new application to market. Texas Instruments has developed a series of software and hardware tools, as well as a user help and information system. Some of these tools offer an integrated hardware and software environment: IDE, *Integrated Development Environment.*

## V.l Software development tools

Table 3 shows the DSP software development tools.

| Software development tools | Assembler, link editor - Assembly language tools |
| --- | --- |
| | Compilateur C - *code generation tools - Optimizing C compiler* |
| Simulation and debugging tools | Simulator and DOS or Windows debugging interface |
| | Integrated development environment: *Code composer* |

Tab 3, DSP software development tools

## V.l.l Assembler and link editor

Users can choose between writing their application program in C or in assembler. Writing directly in assembler gives better performance in terms of execution time and memory occupation, but it is more tedious and requires a thorough knowledge of the processor architecture. After writing a program in assembler to obtain an executable program, the program must be assembled and then linked. Assembly language tools create and use object files in the COFF format (Common Object File Format) to facilitate modular programming. Object files contain separate blocks (called sections) of code and data that can be loaded into the DSP memory space.

**- Using assembler**

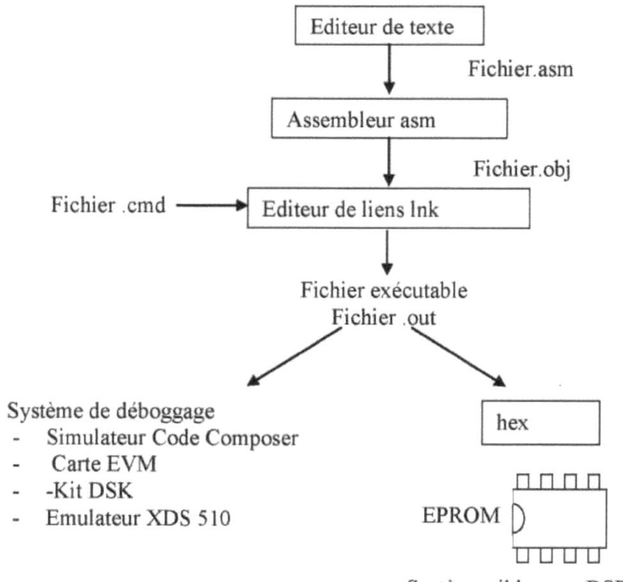


Fig 19.a, Assembler development process

**- Using the C compiler**

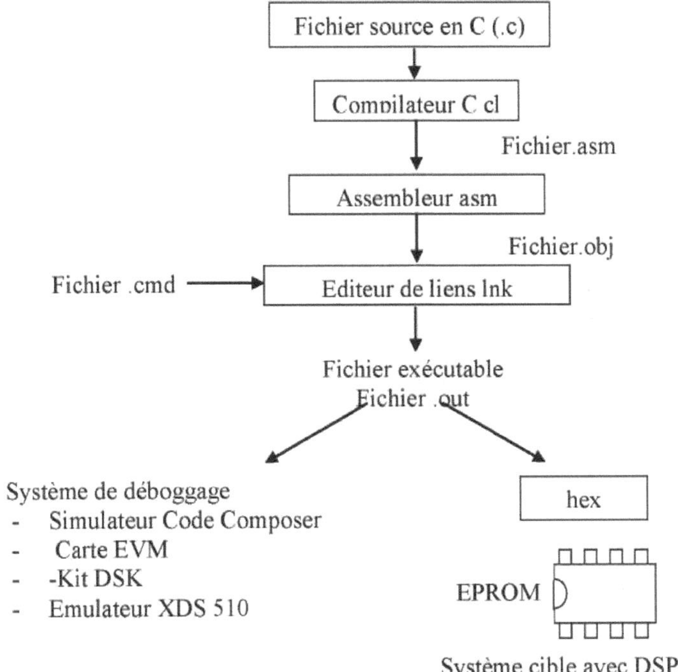

Fig 19.b, C compiler development process

## V 1.2 Simulator and DOS or Windows debugging interface

The simulator can be used to simulate the execution of a program written in C or assembler for a DSP in the TMS320 family, enabling a program to be verified without the real-time aspects. Main features of the simulator :

- Simulation of DSP instructions on a host computer (PC for example) at C or assembler level.
- Configurable, multi-window Windows GUI with drop-down menus and on-line help. The main windows are the C and assembler program windows, the DSP registers window, the memory windows and the variable display windows.
- Visualisation of the C and/or assembler program
- Viewing, modifying and initiating programmes
- Possibility of defining and modifying a memory card accessible by the DSP Viewing and modifying DSP registers
- Visualising C variables
- Automatic update of memory and register contents on the screen after one or more instructions have been executed.
- Running the programme step by step
- Measurement of the number of time cycles required to execute instructions to assess speed performance
- Possibility of setting or deleting programmable breakpoints on acquisition, reading or writing to

27

memory, data values on the bus, or error conditions
- The ability to monitor the values of the accumulator, programme counter or auxiliary registers
- A dynamic *profiling* and analysis model to analyse programme performance and identify critical areas for optimisation
- Ability to connect a file to an I/O port, simulating I/O such as reading or writing to an asynchronous serial port
- Ability to simulate external interruptions

## V.2 Integrated development environment

Code composer is an integrated development environment. Common to the various Texas Instruments DSPs, it can be used on its own or with a hardware target operating in real time, such as an evaluation board.

This environment allows you to build a project, edit files, compile or assemble them, edit links, test and debug a single or multiprocessor application in simulation or in real time on a hardware target. The hardware targets supported are EVM evaluation boards, DSK development kits, as well as Texas Instruments emulators, etc.

The Code Composer simulator allows the application to be tested without the constraints of real time. Code Composer has all the capabilities of the simulator and the debugging interface described above.

## V.2.1 Main functions and possibilities of Code Composer

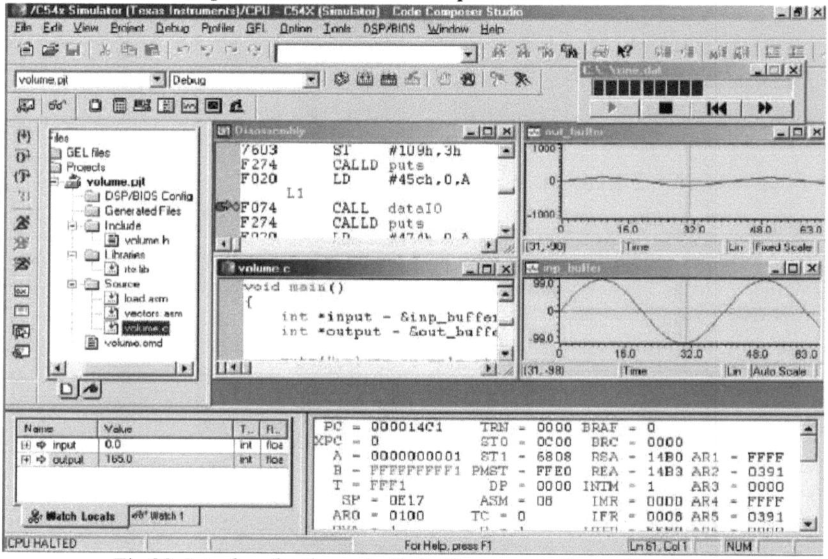

Fig.20, Interface for the main Code Composer software functions

The main window can contain a number of specialised windows:
- A *Dis-Assembly* window containing the program's assembly code
- A window for editing a C source file called volume
- A *Watch* window for displaying variables
- A *Project View* window showing the structure of the current project

- A C54x Registers window displaying the contents of the DSP registers, which can be modified by clicking on them in the window.
- Two graphics windows

There are other windows, such as those displaying memories, and the *CallStack* window, which shows the various function calls that are being executed.

**Project management**
- Creating a project file
- Compiling, assembling and editing links
- Integrated editor

**Integrated debugging**
- Loading a programme
- Break points
- *Run* command
- Free execution, *run free* command
- Step-by-step execution
- Animation, *animate* command
- Stop execution, *halt* command
- Probe points
- Resetting the processor
- Variable display window, *watch* window
- Opening a *watch* window
- Adding a variable to a *watch* window
- Associating files with input/output
- Plots and graphics

**Creating a project file**

Launch Code Composer

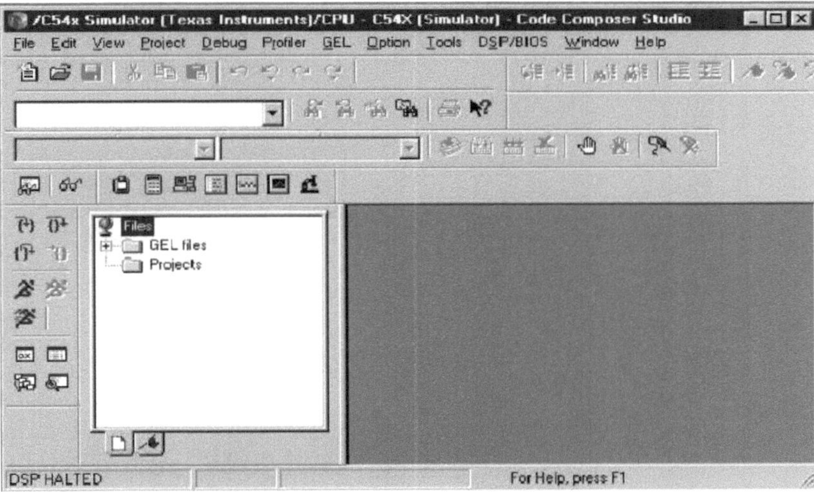

Fig.21, Code Composer launcher interface

Create a new project

29

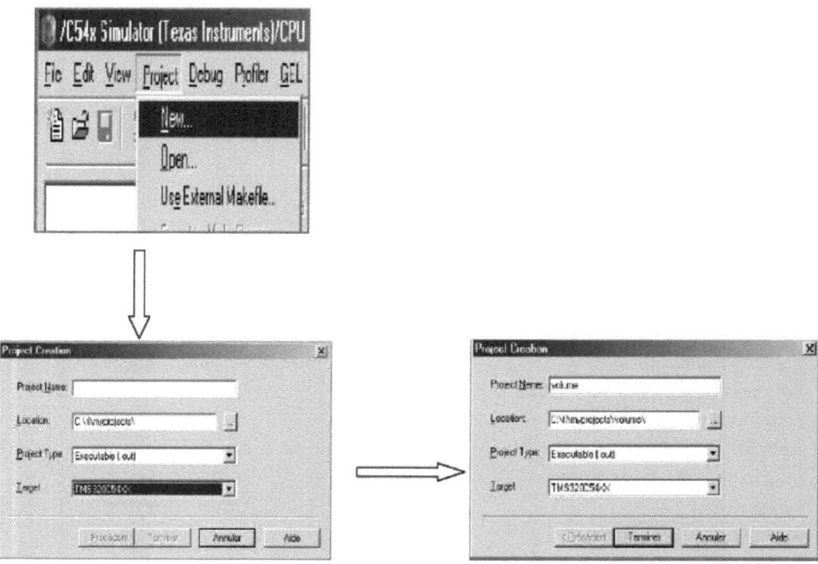

Fig.22, Creating a new project

We add files to the project

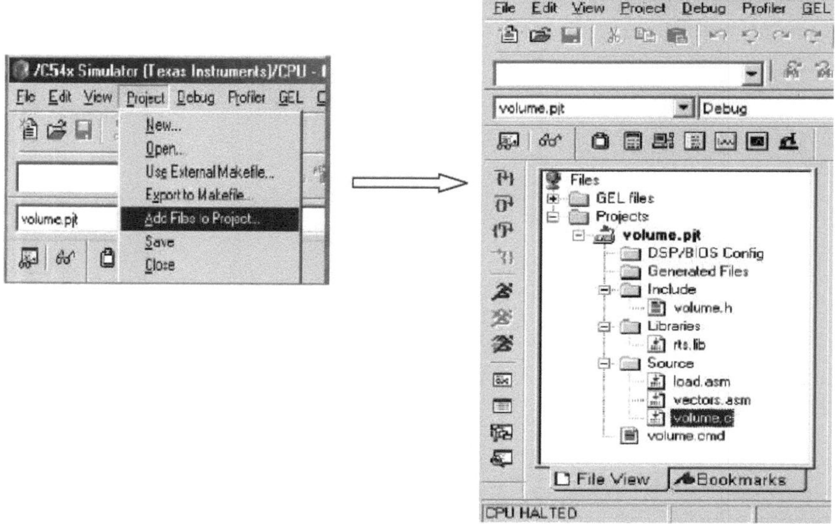

Fig.23, Method for adding files to the project

The project is compiled using the *Project-Build* command

**Compiling, assembling and editing links**

30

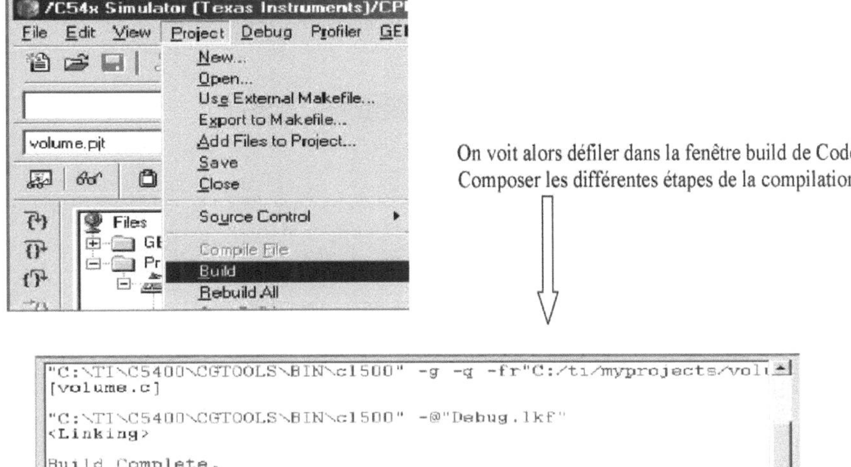

On voit alors défiler dans la fenêtre build de Code Composer les différentes étapes de la compilation

Fig.24, compilation stages

The build window displays the successive compilation, assembly and linking stages. If the compilation, assembly and link-editing stages are completed without error, an executable file (in our case volume.out) will be created in the C:\ti\myprojects\volume\Debug directory and we will be able to test and debug it.

There are also 3 icons that allow you to compile in the following ways

Compile File

Incremental build (only compiles files that have been modified since the last compilation)

Rebuild All

**Integrated editor**

To create and edit a new file (e.g. assembly file (.asm), C file (.c) or command file (.cmd)), use the *File-New* command, which opens a blank editing window. Enter the text and save the file.

Fig.25, interface for creating a new file

To edit an existing file, use the *File-Open* command.

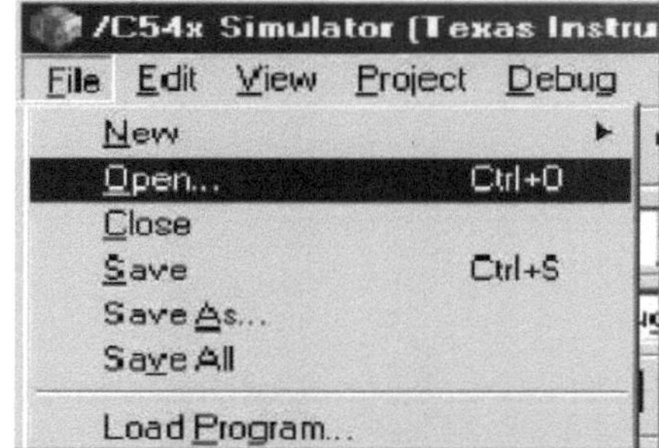

Fig.26, interface for editing an existing file

**Integrated debugging**

**Loading a programme**

To test the volume.out program, first load it with the *File-Load Program* command

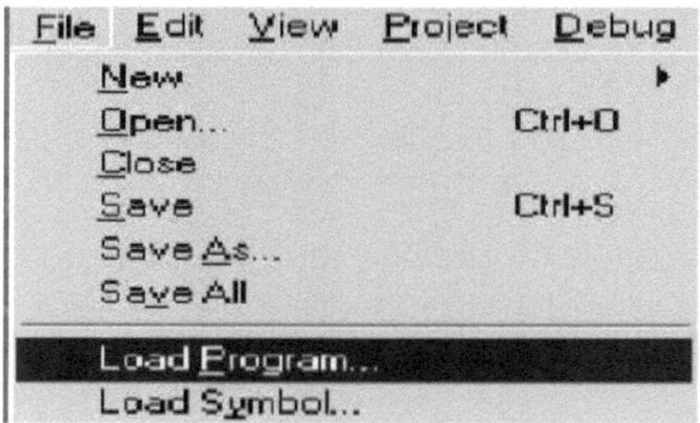

Fig.27, Loading a programme

A *Dis-* Assembler window

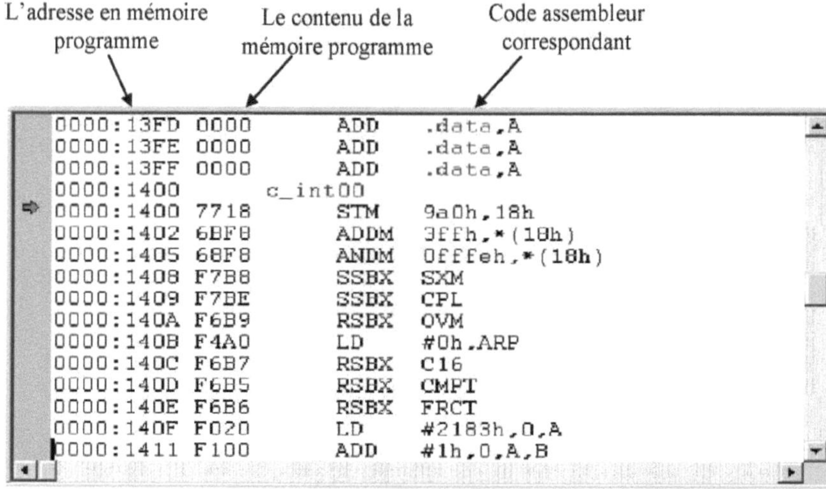


Fig.28, A *Dis-* Assembler window

*Break* **points**

Breakpoints are used to stop the execution of a program when particular instructions are read or when a condition is met.

To place a break point on an instruction, position the cursor in front of the desired instruction in the C or assembler source file, then click on the ***toggle break point*** icon represented by an open hand.

 **Toggle Break point**

When the program is executed, it will stop just before executing the instruction on which the breakpoint has been set.

To delete a break point, click on the ***toggle break point*** icon again.

*Run* **command**

You can start a program by clicking on the *Run* icon, represented by a running blue man. The program runs until it reaches the first stopping point.

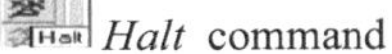 *Run* command


**Free execution, *run free* command**

The *Run free* command can be used to start execution of a program by asking it to ignore all the breakpoints in the Debug menu. You can stop execution with the *Debug-Halt* command.

*Halt* command

**Step-by-step execution**

You can run the program step by step by clicking on one of the three run icons corresponding to the three commands *Debug-StepInto*, *Debug-StepOver* and *Debug-StepOut*. *Debug-StepInto*: executes the instructions one after the other.

*Debug-StepOver:* executes instructions one after the other, treating a function call as a single instruction.

*Debug-StepOut*: executes instructions one after the other, and if activated inside a sub-program, exits the sub-program in a single step.


**Animation, *animate* command**

Use the Debug-animate command or the icon to start a run. When the program hits a breakpoint, the simulator updates all the windows and then resumes execution.

**Stop execution, *halt* command**

Use the *Debug-Halt* command or the icon to stop a program from running.

**Probe points**

Probe points are like breakpoints. They are associated with either a window or a file. When the program reaches a probe point, execution is not interrupted, but if a window is associated with the probe point, it is updated and if a file is associated, it is accessed to read or write a sample in the file.

This can be added using the *Debug-Probepoint* command or by clicking on the icon : 

**Resetting the processor :**

The *Debug-Reset* command stops execution and initializes all the registers to the state they were in when the DSP was powered up.

**Variable display window, *watch* window**

The *watch* window is used to display variables during application testing.

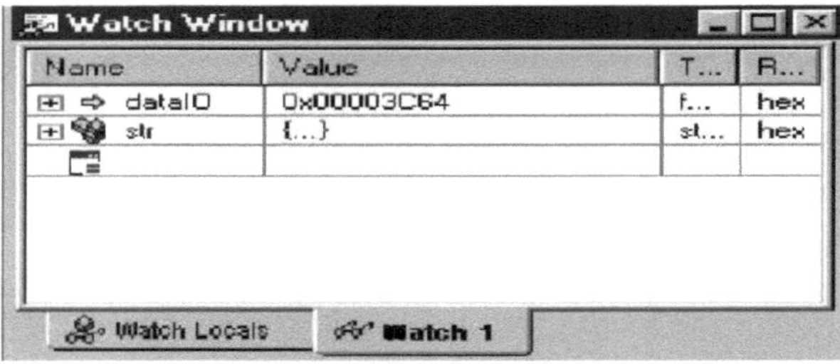


Fig.29, *Watch* window

Opening a *watch* window
You can open window *watch* using the command

*View-Watch Window* or by clicking on the icon : 
**Adding a variable to a *watch* window**
To add a variable to a *watch* window that is already open, you can select the name of the variable containing a C source file or in the Dis-assembly window and then either click the mouse button and select *Add to Watch*

*Window, or on the*

**Associating files with input/output**
Input/output files can replace inputs/outputs such as analogue/digital and digital/analogue converters.
**Plots and graphics**
You can add a graphical window and plot a signal using the commands: *View-Graph-Time/Frequency:* to plot signals or their Fourier transforms.
*View-Graph-Eye-Diagram*: to draw eye diagrams.
*View-Graph-constellation*: to draw constellations.
*View-Graph-Image*: to plot an image (for video processing, for example)

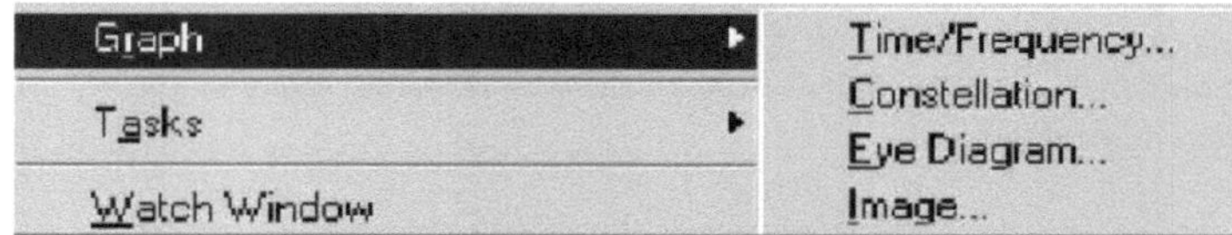


Fig.30, interface for plotting a signal

# REFERENCES


[1]  Y. Bajot, H. Mehrez, Digital Signal Processing Systems, LIP6-ASIM Report, 2001.
[2]  D. Menard, Méthodologies de conversion automatique en virgule fixe pour les applications de traitement du signal, ENSSAT - Université de Rennesl, 2003.
[3]  Jean DEMARTINI, Digital Signal Processors, Pipeline, VLIW, Superscalar: modern architectures, ESINSA 4, 2004-2005.
[4]  Tom Heinrich, DSP processors overview and comparison, 2001.
[5 ] E. Ifeachor, DSP processors and dsp implementationl, 11 March, 2003.
[6]  P. Lynn, Introductory Digital Signal Processing, Wiley press, 1998, www.wiley.com.
[7]  DSP16xxx Targets Communications Apps New Lucent Design Extends Conventional Techniques to

Improve Performance, 1997.

[8]   BDTI, Choosing a DSP Processor, Berkeley Design Technology, Inc; 1999.

[9]   Analog Devices' User's Manual ADSP-2100, 1986.

[10]   C. Moerman, Instruction Sets in DSP Architectures, proc. of the ICSPAT99, Orlando FL, 1999.

[11]   J. Sweeney, Superscalar Techniques Applied to Digital Signal Processing, proc. of the ICSPAT98, Toronto, Canada, 1998.

[12]   O. Wolf, J. Bier, TigerSHARC Sins Teeth into VLIW, Microprocessor Report, December 1998.

[13]   BDTI, DSP Software Optimization Techniques for the Latest Processors, presented at the Embedded Systems Conference (ESC), September 1999, www.bdti.com.

[14]   J. Fridman, Z. Greenfield, the TigerSHARC DSP Architecture, IEEE Micro, Jan 2000.

[15]   R. Grehan, J. Bier, J. Eyre, Massana Unveils DSP Coprocessor Core, Microprocessor Report, November 1999.

[16]   J. Eyre,J. Bier, Infineon's TriCore Tackles DSP, Microprocessor Report, April 1999.

[17]   panorama des processeur du traitement du signal.WWW.lip6.fr /reports.

[18]   Texas Instruments, TMS320C10, digital signal processor, 1986.

[19]   Texas Instruments, TMS320C11, digital signal processor, 1986.

[20]   Texas Instruments, Code Composer Studio Getting Started Guide, Literature Number: SPRU509C, November 2001.

# I want morebooks!

Buy your books fast and straightforward online - at one of world's fastest growing online book stores! Environmentally sound due to Print-on-Demand technologies.

Buy your books online at
**www.morebooks.shop**

Kaufen Sie Ihre Bücher schnell und unkompliziert online – auf einer der am schnellsten wachsenden Buchhandelsplattformen weltweit! Dank Print-On-Demand umwelt- und ressourcenschonend produziert.

Bücher schneller online kaufen
**www.morebooks.shop**

Printed by Books on Demand GmbH, Norderstedt / Germany